はしがき

　このＬＥＤ照明の星座の表現は、星座のイラストを絶句で４つに分けて構成されたものであり、そのストーリ等は星座の内容により表現が異なる。例えば、乙女座の場合は、二人の女性が一人になるまでの過程で、一人の女性の畝が麦穂を持ち、もう一人の女性の葉が、ナツメヤシの葉を持つ、この部分のところが第二の句「承」または第三の句「転」で表現されるように著作権企画を行ったものである。

　1980年代の初めに発明を4コマ漫画で表現した出版、イラストを絶句で表現した出版と歩んでまいりましたが、今回はギリシャ神話の星座とLED電球を関連させ、星座の輝きを深めたものである。

編集部

Preface

Expression of the constellation of this LED lighting consists of Chinese quatrains by dividing into four in the illustration of a constellation, and that story differs in expression according to the contents of the constellation.

For example, in Virgo, it is a process until two women become one person, The copyright plan was performed so that this portion in which a one woman's Furrow has an ear of wheat, and another woman's Frond has a leaf of Palm tree might be expressed with the second phrase "accept" or third phrase "change."

Although it followed with the publication which expressed invention by the four-frame strip cartoon early in the 1980s, and the publication which expressed the illustration by the Chinese quatrain, the constellation and LED electric bulb of Greek mythology are related, and brightness of a constellation is deepened this time.

Editorial department

目　次

１、ＬＥＤ照明　星座の世界（イラスト解説）

(1)　乙女座--5

(2)　ふたご座--6

(3)　かんむり座--7

(4)　琴座--8

(5)　白鳥座--9

(6)　牡羊座---10

２、English of the usage

LED lighting　　The world of a constellation　　（illustration）

(1) ---11

Virgo

LED lighting with which two persons became a one goddess

(2)--12

Gemini

LED lighting with which twins' God and a younger brother made his elder brother deathless

(3)--13

Corona Borealis

LED lighting which makes a monster eradicate

(4)--14

Lyra

LED lighting with which the tone of a koto sounds

(5)--15

Cygnus

LED lighting which flaps a night sky

(6)--16

Aries

LED lighting which illuminates flight

3、宣伝・使用方法--17

4、公報解説--18

5、Patent journal English --41

　　　あとがき--47

1、ＬＥＤ照明　星座の世界　（イラスト解説）

⑴　乙女座

二人が一人の女神になったＬＥＤ照明

二人が一人になった夜空の女神。

夜空の一人の女性、畝の星座と別な女性、葉の女性の星座がいた。

畝の星座の女性は、麦の穂を持ち、葉の女性の星座は、椰子の葉を持っていた。

二人の女性が一人になり、輝いた女神となった。

⑵　ふたご座

双子の神、弟が兄を不死性にしたＬＥＤ照明

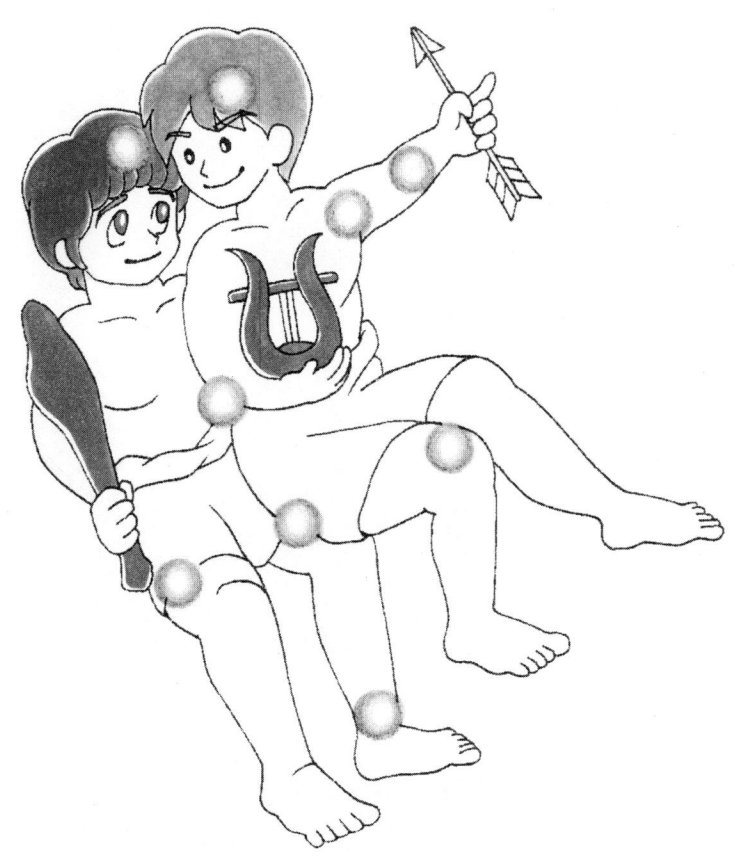

夜空に二つの仲良し星座。

二人の星座が肩を並べている。

一人の星座の弟はポリュデウケースで、不死の神であり、別な星座の兄はカストールで人間であった。弟は兄に不死性を与えた。

二つの星座の双子は、神として夜空で半分過ごし、昼間は人間として地上で過ごしている。

(3) かんむり座

怪獣を退治させるＬＥＤ照明

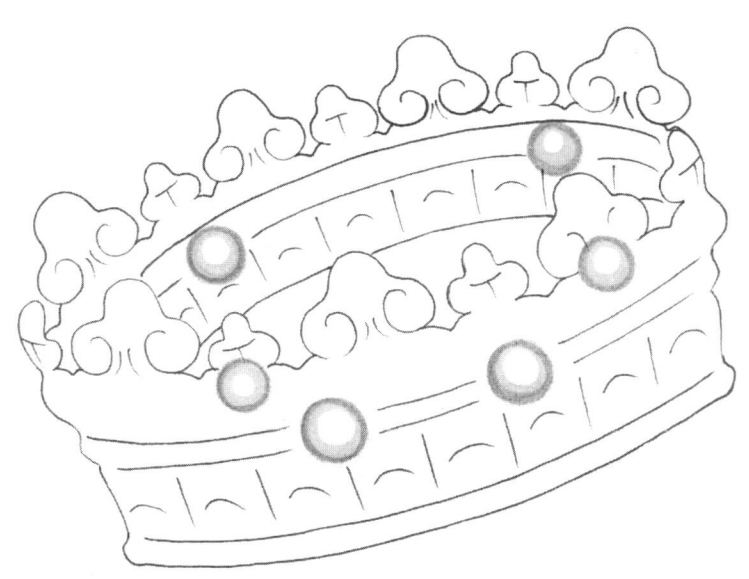

クレタ島の怪獣を退治した勇士。

クレタ島の王様ミーノースは、怪獣ミーノータウロスに毎年、少年・少女を生贄に捧げていた。

王様は怪獣を退治させるために生贄に勇者テーセウスを忍ばせた。王様の娘王女アリアドナエアは、勇者と恋に落ちて島を脱出するが、離れ離れになり、王女は悲しむ。

酒の神ディオニュソスが王女を妻に迎え、冠を与えた。

⑷　琴座

琴の音色が響くＬＥＤ照明

父親から琴を譲りうけたオルペウスは音楽家となった。

オルペウスはエウリュディケーと結婚し、妻となった。

妻エウリュディケーは、毒蛇に噛まれ死んだ。夫オルプウスは妻を戻すために琴を弾き、冥神ハーデースに頼んだ。　琴の音色が良いので妻を戻すことを許すが、帰る途中、後ろを振り返ってはいけないことを告げられる。しかし、家の手前で振り返ってしまい。妻は連れ戻された。

オルペウスは妻が永遠に戻れないので、川に身を投げて死んだ。流れている琴をゼウスが拾い琴座となった。

⑸　白鳥座

夜空を羽ばたくＬＥＤ照明

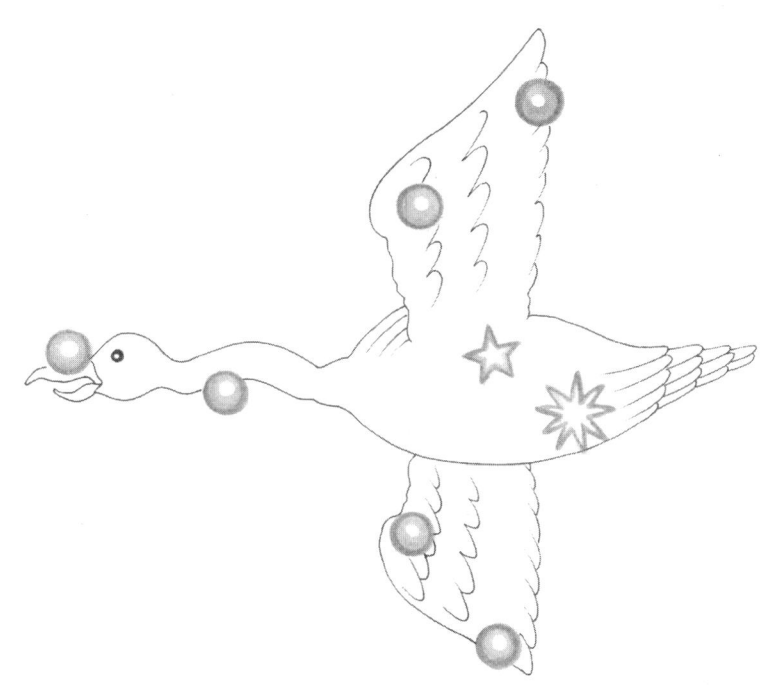

大神ゼウスは王妃レダに恋をする。

大神ゼウスは白鳥に化ける。

大神ゼウスは王妃レダに接近し、王妃レダは卵２個を産む。

２個の卵が双子座の兄弟となった。

⑹　牡羊座

逃亡先を照らすＬＥＤ照明

羊に乗り、継母の生贄から逃れる。

ボイオティア王国に王様の息子プリクソスと双子の妹ヘレーがいた。

二人は継母に生贄にされる運命であったが、これを知った主神、ゼウスが二人を乗せる羊を遣わせ逃げた。妹ヘレーは途中、羊から落ちて海で死んだ。

王様の息子プリクソスは逃亡先の王様に生贄にした羊の金の皮を差し上げた。

2、English of the usage

LED lighting   The world of a constellation   (illustration)

(1)

Virgo

LED lighting with which two persons became a one goddess

The goddess of the night sky where two persons turned into one person.

The constellation of the one woman, and a woman different from the constellation of a "furrow" and the woman of a leaf was in the night sky.

The woman of the constellation of a furrow had an ear of wheat and the constellation of the woman of a leaf had a leaf of a palm tree.

Two women became one person and became the goddess which shone.

(2)

Gemini

LED lighting with which twins' God and a younger brother made his elder brother deathless

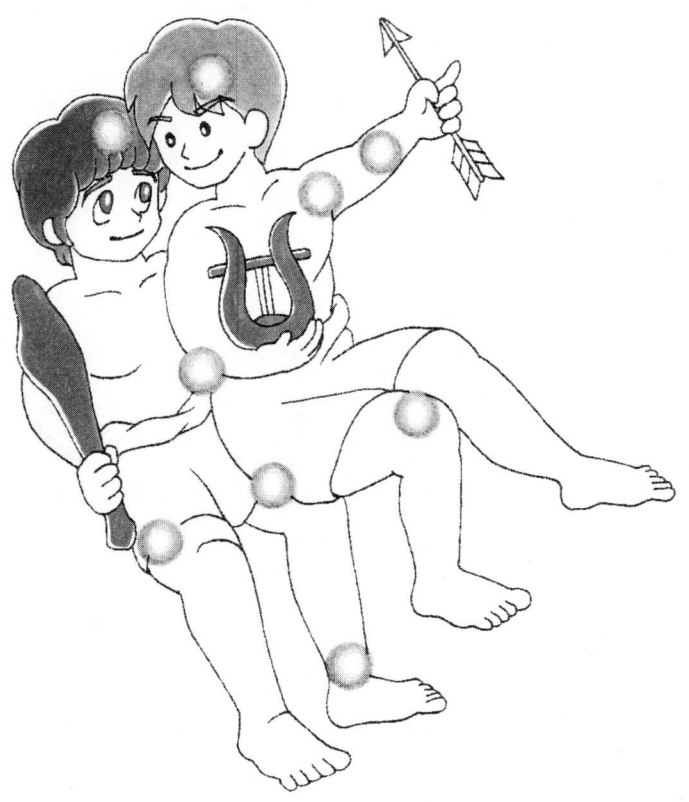

They are two good friend constellations to a night sky.

Two persons' constellation is putting the shoulder in order.

Younger brother Polydeuces of one constellation was God of deathless, and elder brother Castor of another constellation was man.
The younger brother gave his elder brother deathless.

The twins of two constellations are living on the ground as man daytime.
It became God and is living in the universe night.

(3)

Corona Borealis

LED lighting which makes a monster eradicate

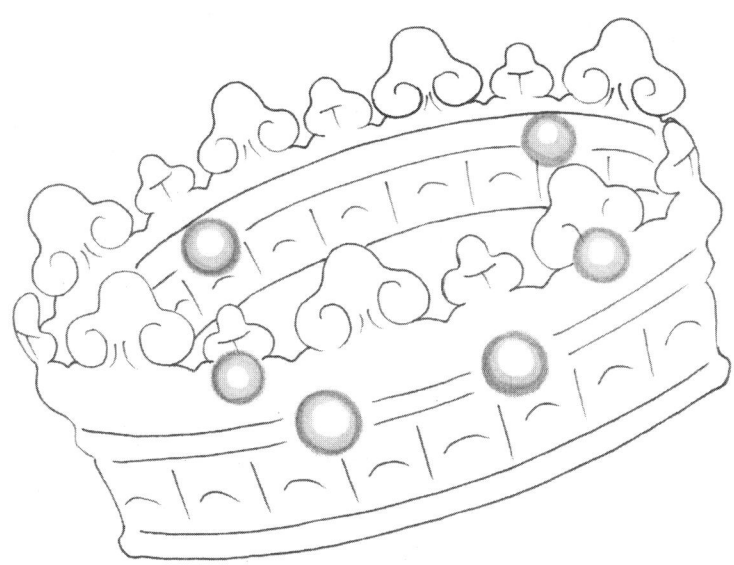

The warrior who eradicated the monster of Crete

King Minos of Crete offered the boy and the girl to monster Minotaur every year at sacrifice.

Since a king made a monster eradicate, he made Sacrifice bear Brave man Theseus.
Although daughter princess Alia, a king, falls to love with a Brave man and it escapes from an island, it gets separate and a princess feels sad.

God Dionysos of alcohol made the princess the wife and gave the crown.

(4)

Lyra

LED lighting with which the tone of a koto sounds

Orpheus who got the koto from the father became a musician.

Orpheus married Eurydice and became a wife.

Wife Eurydice was blown by the poisonous snake and died.
Husband Orphee flipped and asked God hades for the koto, in order to return the dead wife.
Since it is good at a koto, it allows returning a wife, but don't see back while returning.
However, since it had seen before the house, the wife was taken back.

Since a wife could not return forever, it was drowned into the river and Orpheus died.
Zeus gathered the koto which is flowing and it was set to Lyra.

(5)
Cygnus

LED lighting which flaps a night sky

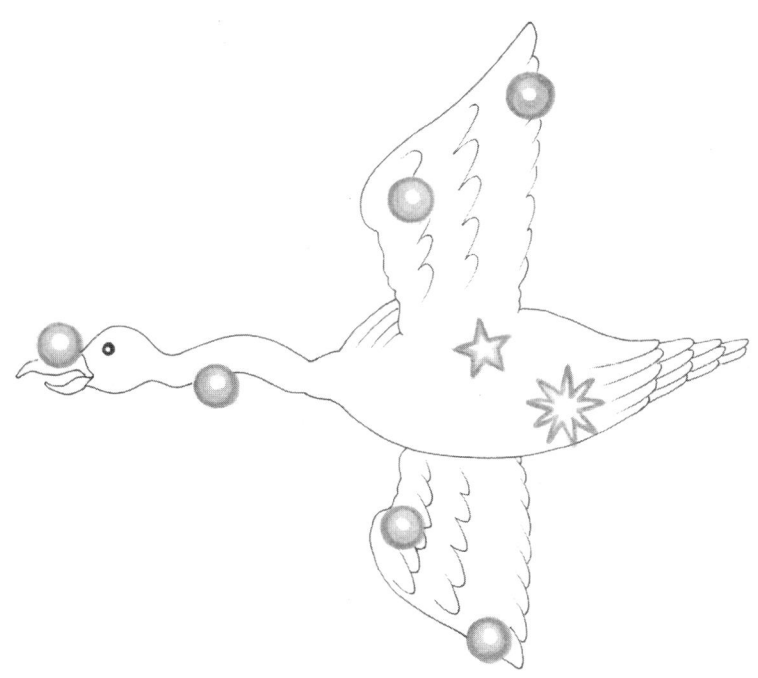

God Zeus is love to queen Reda.

God Zeus Transform oneself into a swan.

God Zeus approaches queen Reda and queen Reda produces two eggs.

Two eggs became a brother of the Gemini.

(6)
Aries
LED lighting which illuminates flight

It rides on the sheep and escapes a stepmother's Sacrifice.

Son Phrixus, a king, and younger sister Helle, twins, were in the Boeotia kingdom.

A stepmother makes two persons sacrifice.
God Zeus who got to know it dispatched the sheep which picks up two persons, and escaped.
On the way, younger sister Helle fell from the sheep and died in the sea.

Son Phrixus, the king, gave the king of the flight place the skin of gold of the sheep made into sacrifice.

３、宣伝・使用方法

１、特徴・効果

　　現代社会の電気消費量の需要・供給が原子力発電に頼っていて、大きな社会問題になっている昨今、この知的財産の果たす役割は大きいものとなっている。

　　ＬＥＤ電球自体の消費電力が八分の一となるので、本件を使用すると各工場、各家庭で消費する電力がメッキリ少ないので、発電がエコにつながる。　家庭での電力使用量が激減するので、その分を工場や会社で使用でき、且つ、二万回の点滅に耐えるので極めて経済的である。

２、点滅・使用方法

　　声センサーにより、点燈・消灯ができるので、使用方法は至って簡単である。

　　色々の明るさは、部屋の広さにより、電球の組み合わせで自由にできる。　多目的使用方法として、⑴玄関、⑵トイレ、⑶寝室、⑷子供部屋、⑸居間、⑹客間など、用途も多目的である。　更に、（イ）天井、（ロ）壁、（ハ）外灯などに取り付けできる。

北原知子

３、公報解説
　　　実用新案登録第３１８１１２６号
　　　考案の名称；多種多様ＬＥＤ電球室内外照明器具
　　　実用新案権者；北原知子

【要約】

【課題】丸い電球型のＬＥＤ電球を複数個、多種多様な形状で配置することにより、一まとまりの照明機器とした室内照明器を提供する。

【解決手段】本体部と蓋部分とを有し、内側に複数のＬＥＤ電球を、下から見て少なくとも三方向、四方向、六方向、または八方向、正方形、六角形、八角形、円形のそれぞれの「型」になるように数個づつ普通の丸いＬＥＤ電球を設置して室内照明機器とする。また、内側に複数のＬＥＤ電球を下から見て少なくとも三角形、四角形、六角形、または八角形の頂点の位置に取り付けて天井吊り下げ式照明機器としても良い。さらには、白鳥座など星座形に該ＬＥＤ電球を配置しても良い。

【選択図】図１－１

【手続補正書】

【提出日】平成２４年１０月２２日（２０１２．１０．２２）

【手続補正１】

【補正対象書類名】実用新案登録請求の範囲

【補正対象項目名】全文

【補正方法】変更

【補正の内容】

【実用新案登録請求の範囲】

【請求項１】

本体部と蓋部分とを有し、内側に複数のＬＥＤ電球を、下から見て少なくとも三方向、四方向、六方向、または八方向、正方形、六角形、八角形、円形の灯りに数個づつ普通の丸いＬＥＤ電球を設置した室内照明機器。

【請求項２】

本体部と蓋部分とを有し、内側に複数のＬＥＤ電球を下から見て少なくとも三角形、四角形、六角形、または八角形の頂点の位置に取り付けた天井吊り下げ式照明器具。

【請求項３】

この普通の丸い電球形のＬＥＤ電球を使用したこの考案は、本体とステンドグラス式のカバーをビスで天井に留める普通の丸い電球形の多種多様ＬＥＤ電球を例えばステンドグラス式星座形の蟹座型、白鳥座型、さそり座型、こと座型、羊座型、ふたご座型、乙女座型、小熊座型、大熊座型、かんむり座型の内側に星の位置に取り付けた星座型式照明器具。

【請求項４】

壁式の本体は金属の管で、大中小のスズランの花の形のグラスの外側を繋げ、各一個ずつ普通の丸いＬＥＤ電球を使用して三段切り替えスイッチ構造にした多種多様ＬＥＤ電球照明器具。

【請求項５】

本体部は立体白鳥形の中に、普通の丸いＬＥＤ電球を数個使用した立体置き型式照明器具。

【手続補正書】

【提出日】平成２４年１１月１日（２０１２．１１．１）

【手続補正１】

【補正対象書類名】実用新案登録請求の範囲

【補正対象項目名】全文

【補正方法】変更

【補正の内容】

【実用新案登録請求の範囲】

【請求項1】

本体部と蓋部分とを有し、内側に複数のＬＥＤ電球を、下から見て少なくとも三方向、四方向、六方向、または八方向、正方形、六角形、八角形、円形のそれぞれの「型」になるように数個づつ普通の丸いＬＥＤ電球を設置した室内照明機器。

【請求項2】

本体部と蓋部分とを有し、内側に複数のＬＥＤ電球を下から見て少なくとも三角形、四角形、六角形、または八角形の頂点の位置に取り付けた天井吊り下げ式照明機器。

【請求項3】

この普通の丸い電球形のＬＥＤ電球を使用したこの考案は、本体とステンドグラス式のカバーをビスで天井に留める普通の丸い電球形の多種多様ＬＥＤ電球照明機器。

例えばステンドグラス式星座形の蟹座型、白鳥座型、さそり座型、こと座型、羊座型、ふたご座型、乙女座型、小熊座型、大熊座型、かんむり座型の内側に星の位置に取り付けた星座型式照明機器。

図面の簡単な説明】

【0009】

【図1】三方形ＬＥＤ多種多様電球室内外照明機器

【図2】四方形 〃

【図３】六方形 〃

【図４】八方形 〃

【図５】円形 〃

【図６】四角形 〃

【図７】六角形 〃

【図８】八角形 〃

【図９】星座かに座ステンドグラス式多種多様ＬＥＤ電球室内－照明機器

【図１０】はくちょう座 〃

【図１１】こと座 〃

【図１２】さそり座 〃

【図１３】おひつじ座 〃

【図１４】かんむり座 〃

【図１５】ふたご座 〃

【図１６】お〻くま座とこぐま座 〃

【図１７】おとめ座 〃

【図１８】壁掛式すずらん形多種多様ＬＥＤ電球室内外－照明機器

【図１９】雪の六角形結晶形多種多様ＬＥＤ電球室内外－照明機器

【図２０】置形式立体スワン多種多様ＬＥＤ電球室内外－照明機器 〔図１〕三方形、設置式、吊り下げ式多種多様ＬＥＤ電球室内外－照明機器。イ、下から見上げた正面図。円形は、ＬＥＤ電球４．９Ｗ３方向に１つづつ。そして四角中央の１ｈは声センサースイッチ点滅器である。ロ、ｇは吊り下げ式のときの吊り下げソケット。ｆは設置式のビス。ハ、ａからｇまでのつなぎ方である。左からａ、ｂ、ｅ、ｃ、ｄ、ｃ、ｅ、ａ、ｂ、ｅ、ｃ、ｄ、ｃ、ｅ、ｂ、ａ、となる。〔図２〕四方形

設置式、吊り下げ式以下同上。イ、〔図1〕と同じ、4方向に1つづつ。〔図1〕と同じ。ロ、〔図1〕のハのようにつづけて、4方向をつなげる。ハ、fのビス。吊り下げ式は吊り下げソケットでとりつけ、とりはずしする。〔図3〕六方形以下同文。〔図2〕のイのようにこれは、6方形で6方向に1つづつ。〔図1〕のハのようにつづけて、中央でコードでわたす。〔図4〕八方形以下同文。〔図2〕のイのようにし、これは八方形で8方向に1つづつ。〔図1〕のハのようにつづけて、中央でコードでわたす。〔図5〕円形以下同文。〔図2〕のイのようにし、これは円形で等間隔に6つ。〔図1〕のハのようにつづけて、円形にコードで渡して点滅するようにする。〔図6〕四角形以下同文。〔図2〕のイのようにし、これは四角形でまわりに9こ。〔図1〕のハのようにつづけて、コードでわたす。〔図7〕六角形以下同文。イ、〔図2〕のイのようにし、これは六角形で角ごとに1つづつ。計6ヶ。ロ、〔図1〕のハのようにつづけて、コードでわたす。〔図8〕正八角形設置式、吊り下げ式以下同文。イ〔図2〕のようにし、これは、八角形で角ごとに1つづつ、計8ヶ。ロ〔図1〕の説明のようにつなげて、コードでわたす。〔図9〕かに座ステンドグラス式多種多様ＬＥＤ電球室内外－照明機器。イ、〔図9〕はかにの部分に色々な色のガラスを使い、ステンドグラス風にする。ロ〔図1〕の説明のようにつなげて、コードでわたす。〔図10〕はくちょう座多種多様ＬＥＤ電球室内外－照明機器。イ、5ヶのＬＥＤ電球を使用。上記の図のように〔図9〕のようにつなげる。ロ、図1のようにつなげてコードでわたす。〔図11〕〔図12〕〔図13〕〔図14〕〔図15〕〔図16〕〔図17〕〔図18〕〔図19〕以上は、今までの説明と同じようにつくる。〔図20〕置形式立体スワン多種多様ＬＥＤ室内外－照明機器。イ、は、置形式立体スワンを強化ガラスで作製し、中に、ビスで接着剤でとめ、ＬＥＤ電球を、aからhまでの素材で、ともるようにつなげる。

充電式でもある。ロはＬＥＤ電球の位置。

【図1-1】

イ.

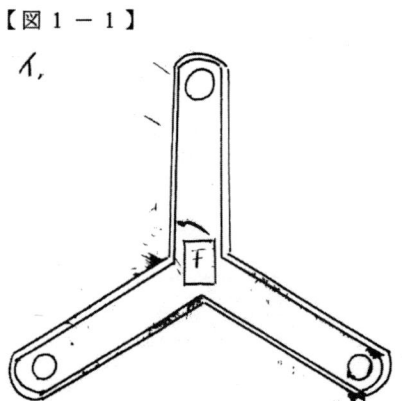

【図1-2】

ロ

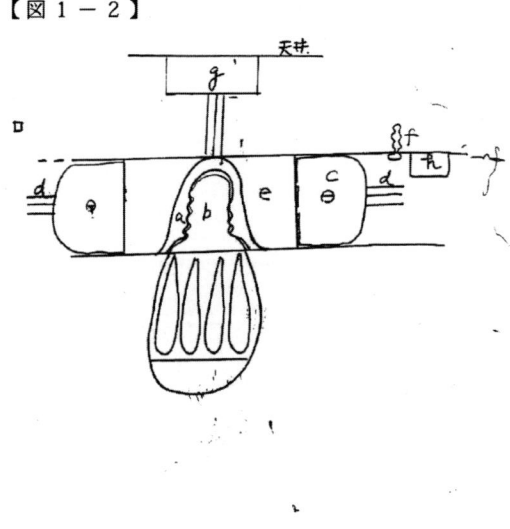

ハ

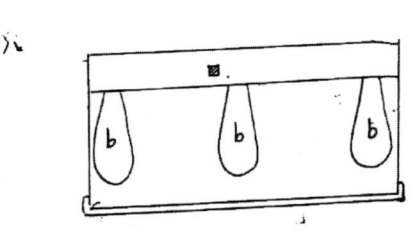

【図2-1】

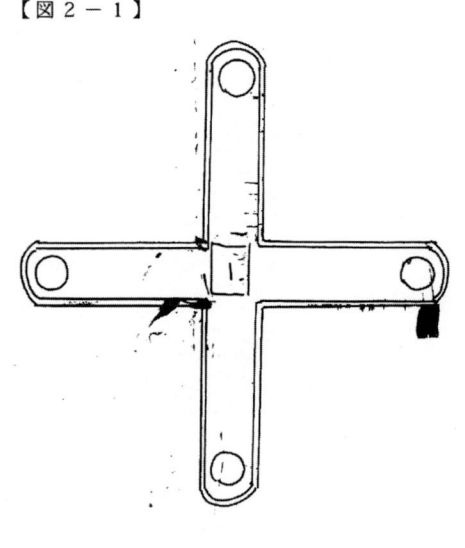

【図2-2】

ロ

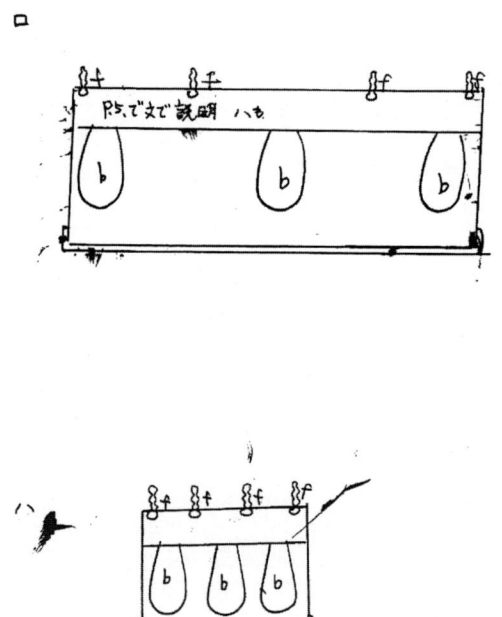

【図3-1】

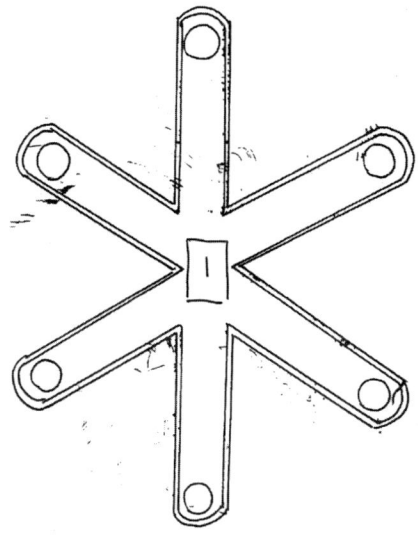

【図3-2】

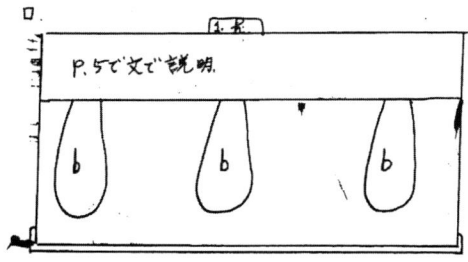

【図4-1】

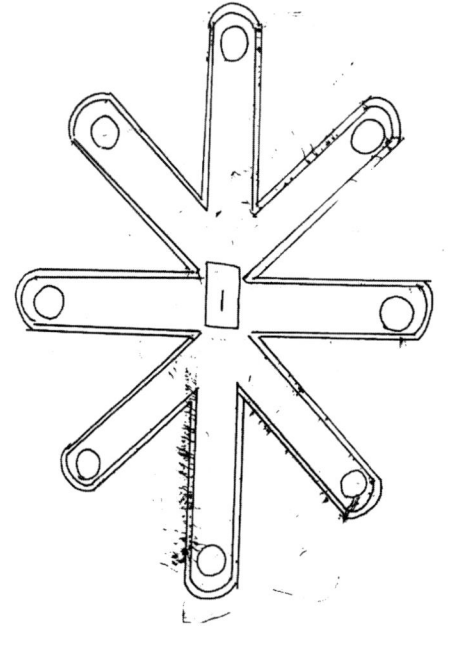

【図4-2】

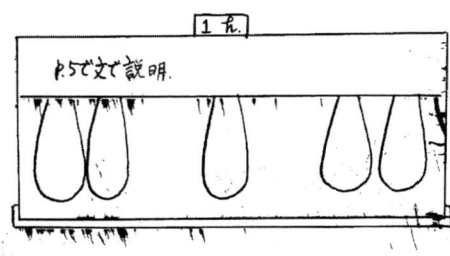

【図5-1】
イ

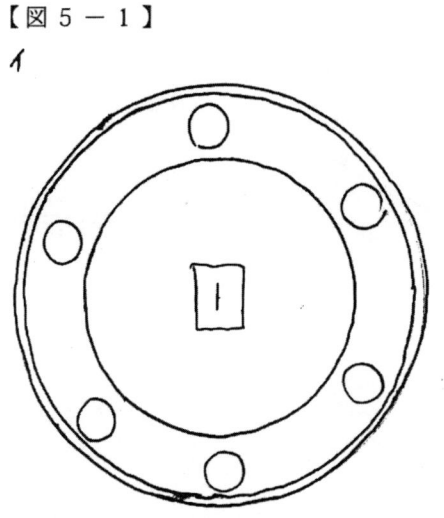

【図5-2】
ロ．

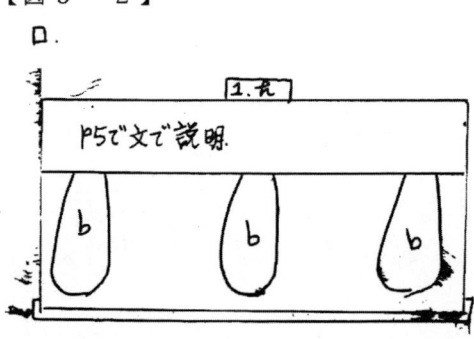

【図6-1】

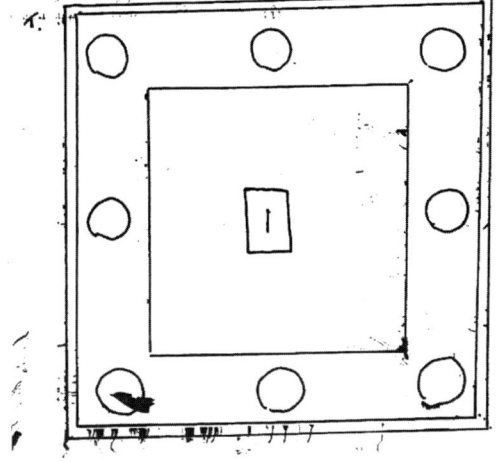

【図6-2】

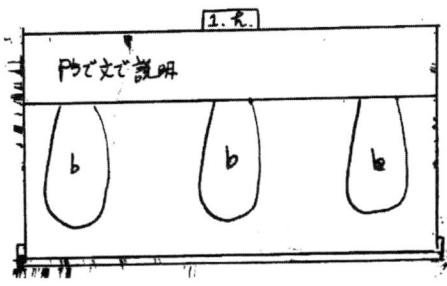

【図7－1】

イ

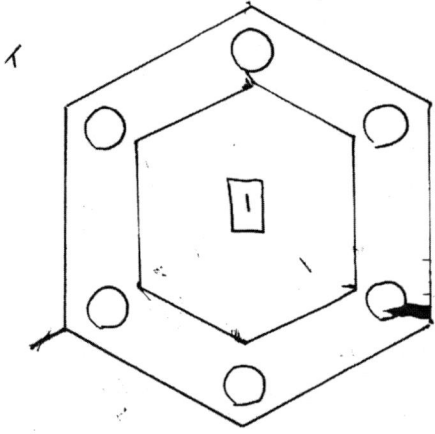

【図7－2】

ロ．

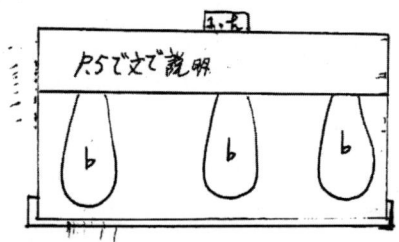

【図8-1】

イ

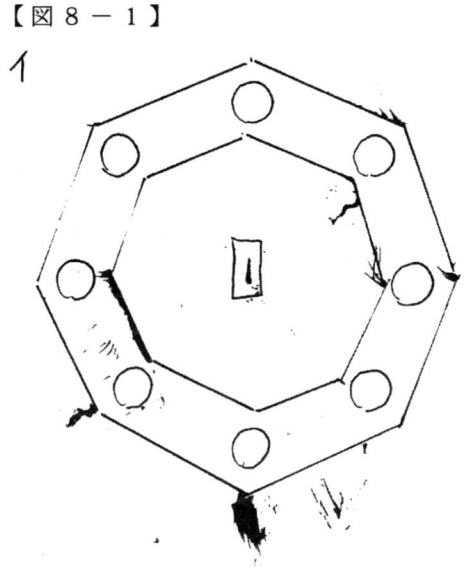

【図8-2】

ロ．

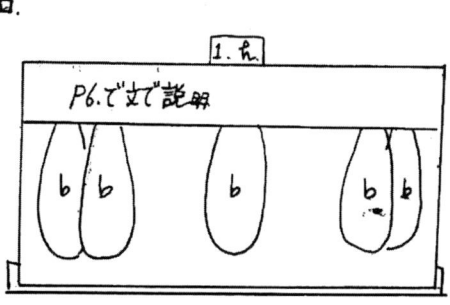

29

【図9-2】

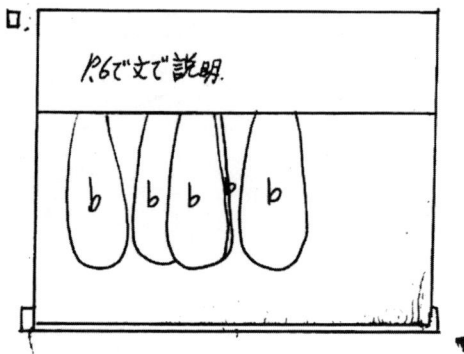

【図9-1】

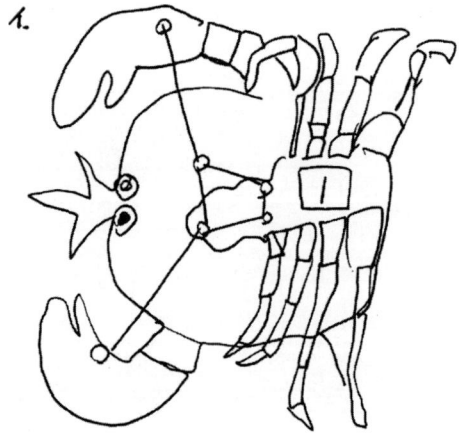

【図10】

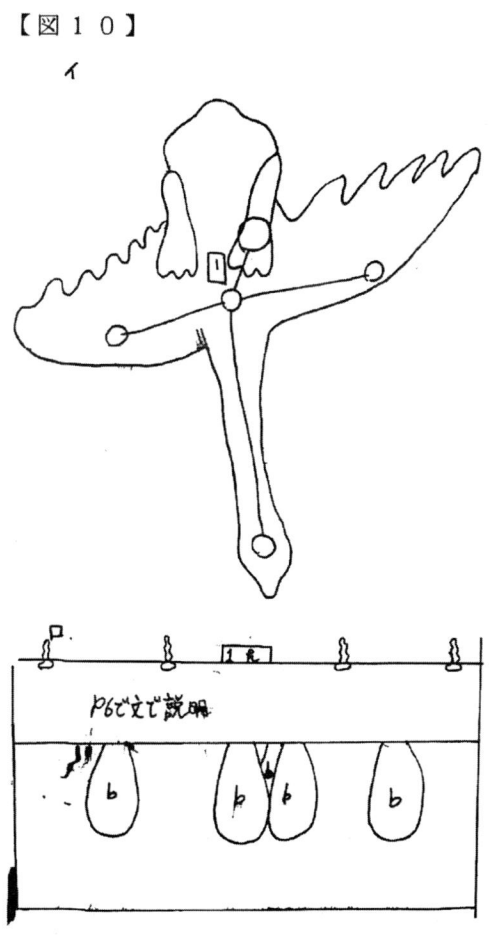

【図１１】

イ．

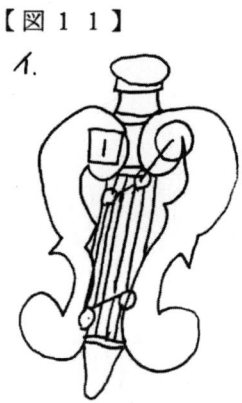

ロ．

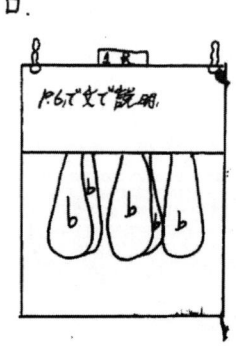

【図12】

イ.

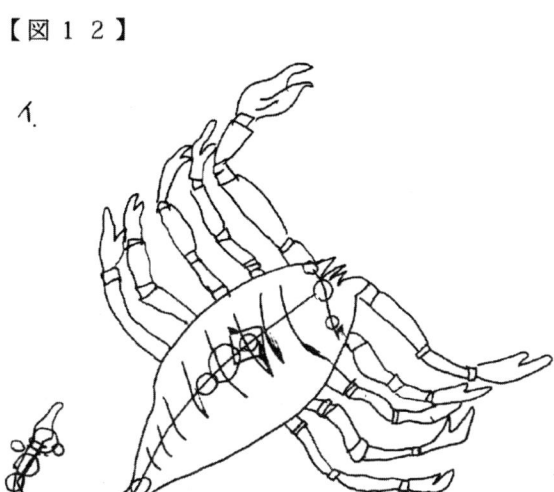

ロ.

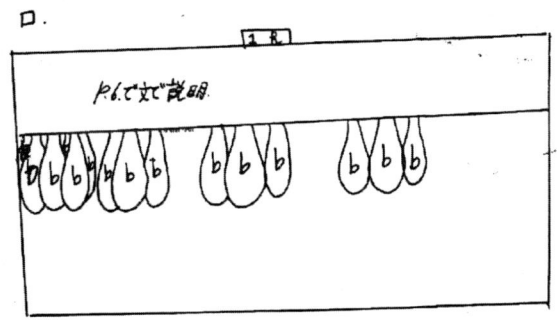

33

【図13】

1.

2.

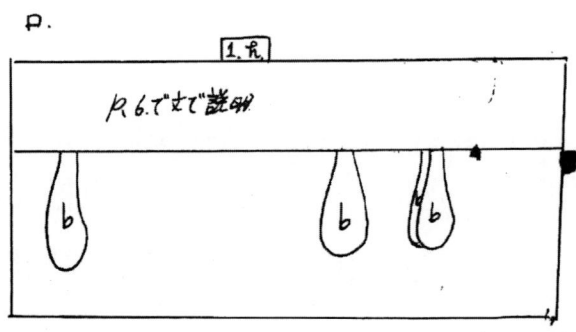

【図14】

イ.

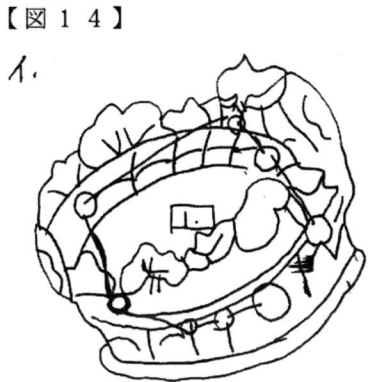

ロ.

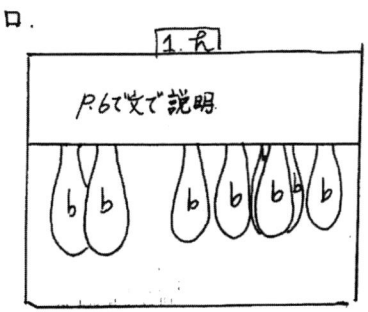

35

【図１５】

イ.

ロ.

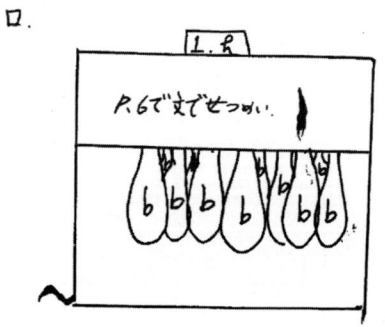

【図16】

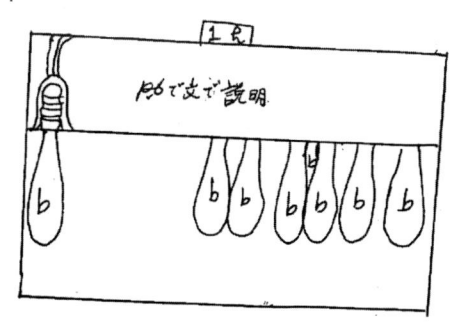

イ.

ロ.

16

図6で文で説明

37

【図17】

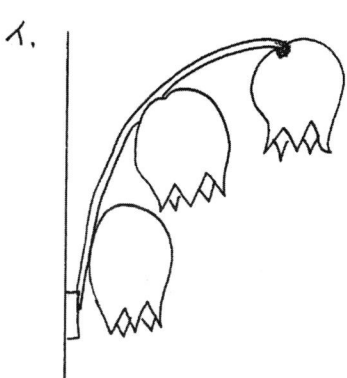

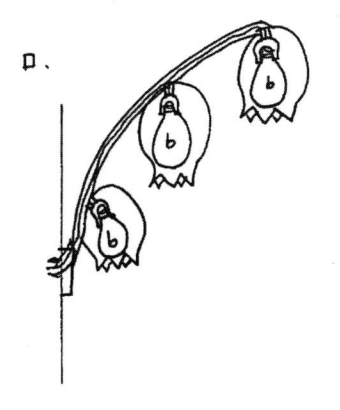

【図18】

イ．

ロ．

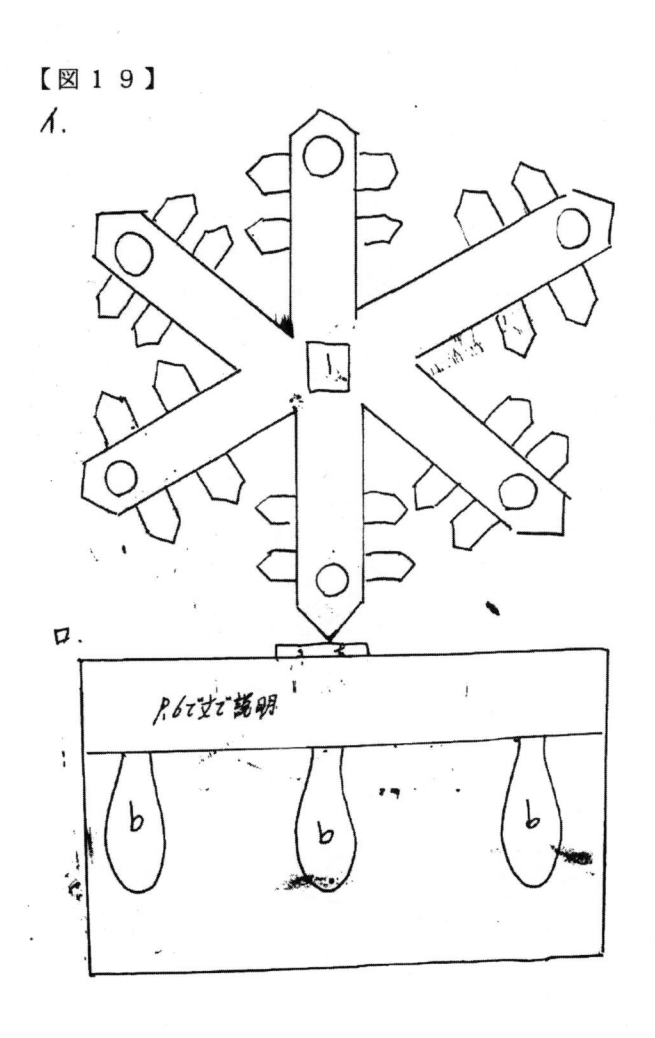

【図19】

イ.

ロ.

P.6で文で説明

5、Patent journal English

DETAILED DESCRIPTION

[Detailed explanation of the device]

[Field of the Invention]

[0001]

This design is related with the mold of a light (interior of a room), and the number of a LED bulb.

[Background of the Invention]

[0002]

It is a new light (for the interior of a room) using some LED bulbs designed in order to make sufficient bright the room of 6-8 . 8-10 . 8-12 mat . at home etc.

[The issue which the outline of a device tends to solve]

[0005]

From the interior illumination instrument by the present, it is one decrease [of 20% of . economy] . LED bulb.

[0006]

It attaches and is 1.7 times the luminosity of this.

[0007]

It will be connected with . Eko if it. Changes to this.

[Means for solving problem]

this design -- the main part of . LED luminaire itself . -- in order to make some LED bulbs into the just right luminosity of . interior of a room, it was.

made to cooperate and unified.

[Effect of the Device]

It is useful to connect an LED look with . Eko, and to lead to . global warming reduction, since [which saves electric power on . social target, and . and . heat dissipation also like] there is nothing.

[Brief Description of the Drawings]

[0008]

1b. 1. stationary type side view

2 Figure . which looked at and raised ** top and was seen from the bottom -> three centers

Outside Five pieces.

**. 1. Hang and lower and it is a formula side view.

Stationary type.

2b. 1. center One piece

The following outside Five pieces

Portion of the 3rd outside Seven pieces.

** Hang and lower and it is a formula.

2 Same as the above.

3. Hang and lower and it is a formula. 8 - for [for 10 mats] 6-8 mats [for 8-12 mats]

1. Type in all directions 16 piece+1 piece center .10 piece Six pieces

2 Hexagonal formula 18 12 six pieces

3 Mitsumoto type . 18 piece I .8 piece - RO .10 piece Six pieces

[The form for devising]

[0009]

although . opening and closing of the cover plate from the side surface of a LED bulb illumination room internal use device body are done with . screw . -- the material is . translucent tempered glass or . translucent fiber reinforced plastic. opening, closing, . hanging and lowering -- irrespective of a formula . -- some of each LED bulb and . -- the main part shakes among . and is kept from carrying out

Attachment of . LED bulb is enabled.

[Working example]

[0010]

According to . accompanying drawing, . working example is described below.

(2) DRAWINGS

1b. Attach a direct . stationary type LED bulb interior illumination instrument LED look with a screw on heavens.

**. The figure looked up at from the bottom

It hangs and lowers and is a formula.

3The figure which was hung, was lowered and was seen from under a formula. It is as in the figure of a front page.

[0011]

With the light for the LED bulb interior of a room, 5.6.7 hangs the main part of 1, and 1.5.6.7 of the (above-mentioned) is lowered, and becomes a formula and a stationary type.

[0012]

Because the working example of this design is what consists of composition like the above-mentioned.

a LED bulb some -- being able to become as a number of one illumination room internal use instrument decided every, and passing to . Eko is proved.

[0013]

1. LED look main part

2. ** lid part

3. ** screw

4 LED bulb use

5 hanging and lowering -- business -- a type

6 Stationary type

7 Hang and lower and it is a formula code.

(1) Description.

1The object for light . 6-8 mats using the LED electrical and electric equipment (electric bulb) which ends by one 1.7 times the luminosity of subject of this with about 20% of . quantity of electricity now was devised.

2The upper, same LED luminaire for . 10-12 mats.

3b. Fixed type.

**. Hang and lower and it is a mold.

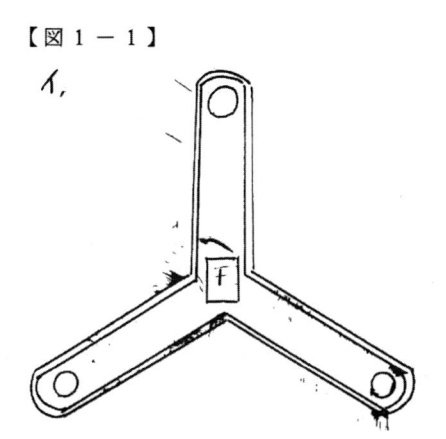

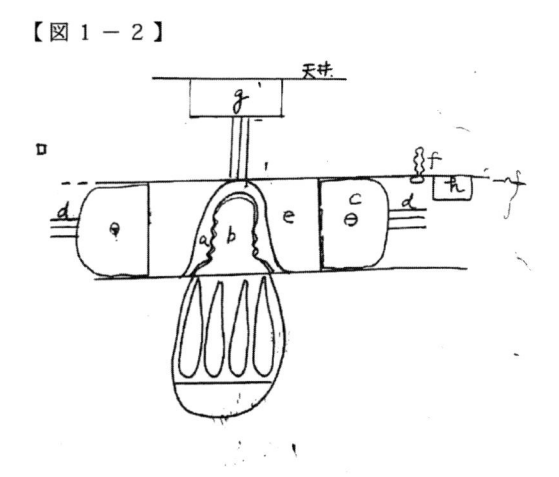

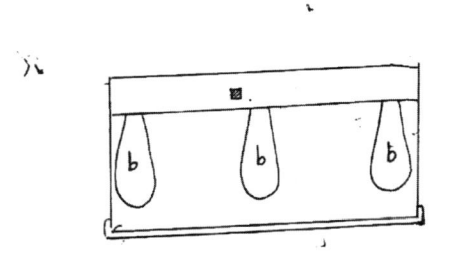

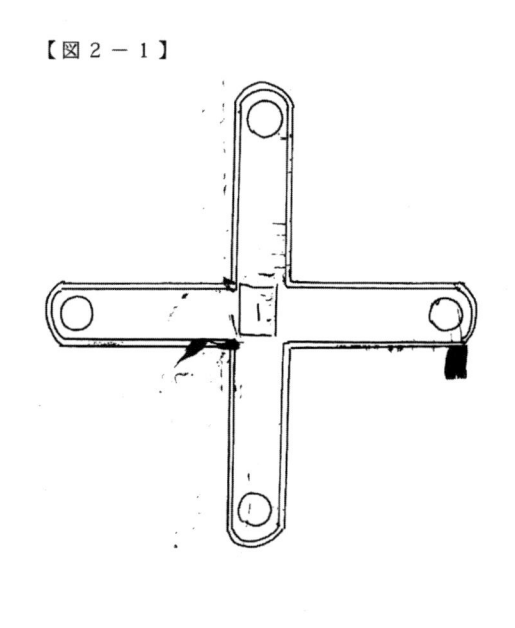

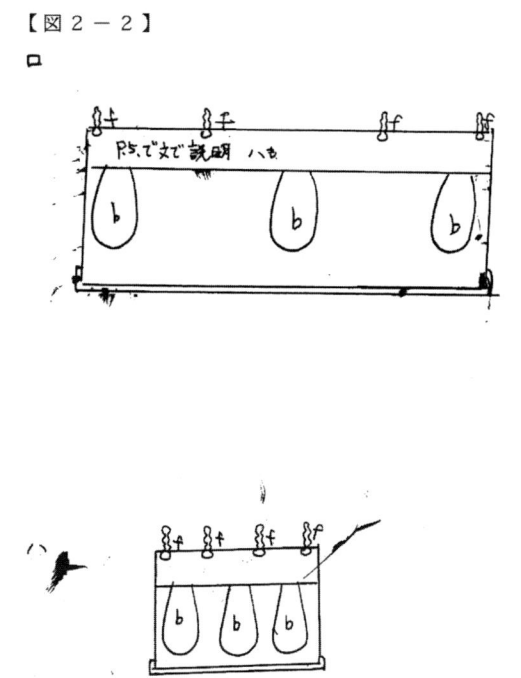

【図1-1】

イ．

【図1-2】

ロ．

ハ．

【図2-1】

【図2-2】

ロ．

ハ．

45

【図3-1】

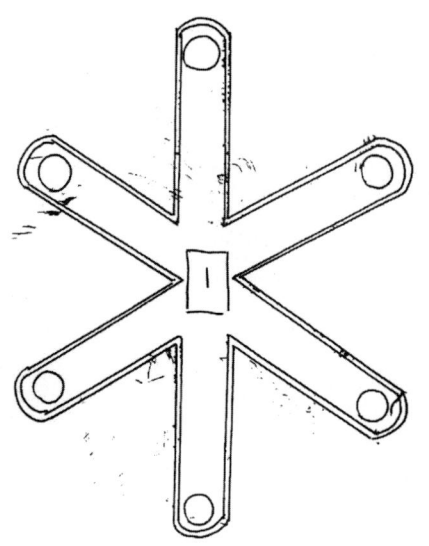

【図3-2】
P.5で文で説明.
b b b

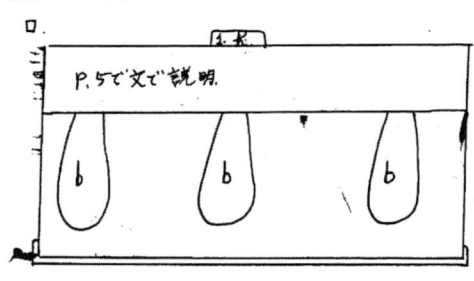

【図4-1】

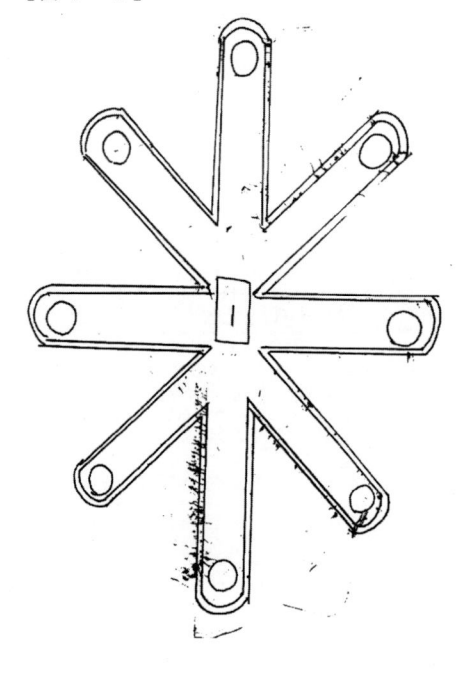

【図4-2】
P.5で文で説明.

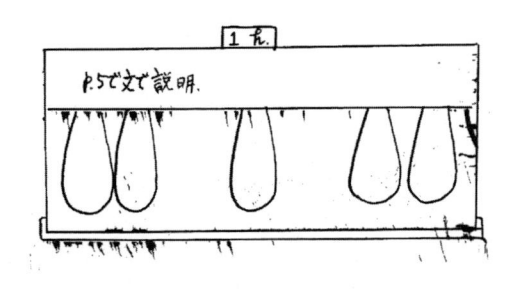

あとがき

　私の「多種多様ＬＥＤ電球室内外照明器具」を発明・考案したのは、ＬＥＤ電球が、電気消費料が普通より、八分の一～六分の一（これは、全体として、大体この位になると考えたので）位になるからです。その上、取替えるにも回数が少なくて（二万回の点滅に耐える）済むからです。

　使い方としては、声センサーがありますので、それを使用しました。

　広い室内には、充分な明るさを考慮して、数個、取り付けるのも可能ですし、組み合わせは自由です。　あと、発電量が少ないので、その必要性のある所や夏暑い国では特に好適です。

著者　北原知子
（きたはらともこ）

LED電球　輝く星座　ギリシャ神話の絶句解説

定価（本体 1,500 円＋税）

―――――――――――――――――――――――――――――

２０１３年（平成２５年）１１月１１日発行

No. KH-018

発行所　発明開発連合会®

東京都渋谷区渋谷 2-2-13

電話 03-3498-0751㈹

発行人　ましば寿一

著作権企画　発明開発連合会

Printed in Japan

著者　北原知子 ©

―――――――――――――――――――――――――――――

本書の一部または全部を無断で複写、複製、転載、データーファイル化することを禁じています。

It forbids a copy, a duplicate, reproduction, and forming a data file for some or all of this book without notice.